VOLUME 12

aero ARMOR-SERIES

UWE FEIST

AERO PUBLISHERS, INC.
329 West Aviation Road Fallbrook, CA 92028

TABLE OF CONTENTS

ISBN 0-8168-2044-9

Library of Congress Catalog Card No. 80-66353

ACKNOWLEDGEMENTS:

Deutsches Museum, München
Thyssen-Henschel, Kassel
Imperial War Museum, London
Bibliothek für Zeitgeschichte, Stuttgart
National Archive
Bundesarchiv, Koblenz
Randolf Kugler
Aero Archive
Archive Uwe Feist

Kampfpanzerwagen "Nb.Fz." (Neubaufahrzeug)

The tank gunnery school in Putlos used three of the Krupp built "Neubaufahrzeuge" to train crews for the emerging Panzertruppe. The 23 ton vehicle was powered by a six cylinder BMW airplane engine of 250 H.P. with a six forward and six reverse gear transmission, which was fully synchronized.

Kampfpanzerwagen "Nb.Fz." (Neubaufahrzeug)

WEIGHT:	19.5 t
CREW:	6
MOTOR:	1 X BMW — 250 PS (HP) 6 cylinder
SPEED:	30 Km/h
RANGE:	265 Km (road)
CAPACITY:	375 Ltr.
LENGTH:	665 cm.
WIDTH:	300 cm.
HEIGHT:	290 cm.
ARMAMENT:	1 X 7.5 cm Geschütz L/23, 1 X 3.7 cm Geschütz L/45 cannon 3 X 7.9 mm MG machine guns
MANUFACTURER:	Krupp (3) and Rheinmetall (2) 1934

"Nb.Fz." number 3, 4 and 5 were sent to Norway in April of 1940 to see some limited action in which the performance of the AFV's was less than expected.

The "Nb.Fz." had a crew of six consisting of the commander, gunner and leader housed in the turret, one gunner each in the front and rear MG turrets and the driver. The vehicles operating in Norway were plagued by constant mechanical breakdowns and the lack of range which was only 120 Km. (appr. 66 miles). This photo was taken on May 15, 1940, near Oslo, Norway.

The "Neubaufahrzeug" was equipped with a 20 Watt radio sender and UKW receiver with a maximum range of 5 Km. The crew used an intercom system. The 2 m antenna is located on the turret side above the escape hatch.

The frontal armor was 20 mm thick, the side 13 mm and floor armor only 8 mm. The main armament located in the turret consisted of a Krupp 7.5 cm cannon with a coaxial 3.7 cm cannon with 80 X 7.5 cm and 50 X 3.7 cm rounds stowed in the fighting compartment. The 3 X MG 34 [one in the main turret, and one each in the front and rear MG turret] had 6,000 rounds available. Note the Boschhorn between driver's and MG gunner's position. On the right side of the turret top, the antenna head.

Front view of the Krupp built "Neubaufahrzeug." The vehicle was 6.65 m long, 2.90 m high and had a width of 2.90 m. It crawled at a low speed of 3.9 Km/h and could reach a top speed of 30 Km/h (appr. 16.5 m.p.h.). The 70 ton "Kingtiger" could reach 41 Km/h top speed!

The suspension of the "Nb.Fz." consisted of ten rubber tired double road wheels with four return rollers. The escape-hatch could only be opened from within.

Right side-view of the Krupp "Neubaufahrzeug" in the Reichswehr color scheme of yellow-red brown and green.

A Panzerkampfwagen II (2 cm) Ausf. A of the 12. Panzerdivision is seen advancing through a Russian village during the early weeks of Operation "Barbarossa," July 1941. Note the MG mounting on the turret.

Panzer II Ausf. A with smoke candles installed on the rear fender, a practice soon to be discontinued. The Panzer II made up for the bulk of AFV's in the Panzerdivisions in the early stages of the Second World War. More than 1,000 vehicles of this type advanced into the Soviet Union during "Barbarossa."

A convoy of the 7. Panzerdivision is headed by a Panzer II A. The 9.5 t vehicle could travel at a top speed of 40 Km/h and had a range of 170 Km. (appr. 93 miles). The motorcycle rider is carrying the famous Lica camera, a type favored by many German soldiers during the war.

Infantry armed with light 7.9 mm MG 34 and the Karabiner 98 K rifle are riding on Panzer II A's. Note the stick grenade, the standard German hand grenade attached to the turret. The gadget on top of the muffler is the smoke candle holder which could be activated from within the vehicle.

Panzer II's with their turrets at one o'clock are moving up to check Russian resistance from the woods ahead, which is holding up a convoy of trucks and other unprotected vehicles. One vehicle which was disabled by the ambush can be seen in the background.

A Panzer II Ausf. A with an MG 34 machine gun at ready is crossing over the standard German army bridge, the Magirus Sturmbrücke. The bridge had a capacity of up to 28 t, which included the Panzer IV, the heaviest German tank in service at the time of the Invasion of Russia, June 1941.

The armor plates are lowered into a position which will enable the crew to be effective against targets in the air. The gun could not be depressed low enough to be effective against ground targets.

The side and rear walls of the structure are constructed of two spaced 20 mm armor plates, giving the vehicle a total height of 2,700 cm.

The following photos show the Flakpanzer 43 in various positions during crew training. One crew member is using the Zeiss Entfernungsmesser [range finder].

The 3.7 Flak 43 had a 360° traverse and 90° elevation and was centrally mounted on a pedestal. The weapon was built by Rheinmetall. A total of 240 units were assembled by the Böhmisch-Mährische Maschinenfabrik, Prag.

The Flak 43 had a muzzle velocity of 275 feet per second and a practical rate of fire of 150 rds./min. (theoretical 250 rds./min.). The weight of the weapon was 2,750 lb. (1,237 Kg.). Combat weight of the Fla.Pz. IV was 25 tons.

The Flakpanzer IV carried 413 rounds of 3.7 mm ammunition in clips of eight rounds each which were fed horizontally from the left of the gun. The wiremesh prevented the ejected empty cartridge from flying into the superstructure.

The Flakpanzer 43 "Möbelwagen" was based on the chassis of the Panzer IV Ausf. H. They were primarily intended to increase the anti-aircraft protection of the Panzer units which were hard pressed by overwhelming Allied airpower following the Invasion. Normandy, June 6, 1944.

The Bug-MG was removed and the opening sealed. The engine performance increased to 272 PS and the total combat weight of the Flakpanzer remained at 25 t [Pz. IV Ausf. H - 26 t].

The seven men crew [five gun crews and two to operate the vehicle] found limited protection behind the 2 X 20 mm armor plates of the superstructure which would be lowered during action.

Only some prototypes of the 2 cm Flakvierling 38 auf Selbstfahrlafette Panzerkampfwagen IV were built. The height of the vehicle (3 meters) prompted the troops to call it "Möbelwagen" (furniture van). The 25 ton Flakpanzer had a crew of five men which found limited protection from small arms fire and splinters behind the 10 mm walls of the structure.

Kampfpanzerwagen "Nb.Fz." (Neubaufahrzeug)

Bergepanzer "PANTHER" (Sd.Kfz. 179)

Schützenpanzer MARDER

2 cm Flak 38 auf Sfl. 38(t) [Sd.Kfz. 140]

2 cm Flak 38 auf Panzerkampfwagen 38[t] [Sd.Kfz. 140]

In 1943 the Böhmisch-Mährische Maschinenfabrik, Prague, Czechoslovakia, began converting the now obsolete remaining Pz. Kpfwg. 38[t] into the first fully tracked self-propelled anti-aircraft tank. Note insignia of the 12.SS-Pz.Div. "Hitlerjugend" on the rear structure.

2 cm Flak 38L/55 auf Sf. 38[t] [Sd.Kfz. 140]

WEIGHT:	9.8 t
CREW:	5
MOTOR:	1 X Praga AC, 6-Cyl., 150 PS [HP], water
SPEED:	42 Km/h
RANGE:	S-210 Km, G-140 Km
CAPACITY:	218 Ltr.
LENGTH:	461 cm.
WIDTH:	213 cm.
HEIGHT:	225 cm.
ARMAMENT:	1 X 2 cm Flak 38 L/55
MANUFACTURER:	Praga 1943-1944 162 built

A 2 cm Flak 38 anti-aircraft gun was mounted at the rear of a 38(t) chassis and enclosed by a superstructure consisting of eight 10 mm armor shields.

The Flakpanzer 38[t] had a combat weight of 9.8 tons and entered service in October 1943. Only 162 vehicles were converted and delivered exclusively to Panzerunits on the Western Front.

Driver's position of a Pz.Kpfw. 38[t]

The 2 cm Flak 38 was introduced in 1940 replacing the 2 cm Flak 30, and was installed with or without a 10 mm gunshield. The gun was fed from the left by a 20 rds. curved box magazine (10 Kg.) of either armor-piercing (A.P.) or high explosive (H.E.) ammunition. Muzzle velocity was 2,950 feet/sec. at a maximum vertical range of appr. 4,000 meters with a rate of fire of 220 r.p.m. (practical) and 450 r.p.m. (theoretical).

The shields were hinged to the bottom and could be folded back to enable the gun to traverse 360°. A Flakpanzer 38(t) of the Flak-Abteilung 12. SS-Panzer-Division "Hitlerjugend" in France, 1944. The crew member is handling a spare barrel.

Leichter Ladungsträger "Goliath" [Sd.Kfz. 302a]

The leichte Ladungsträger [Sd.Kfz. 303a] was an improved version, now powered by a two cylinder 7.5 PS air-cooled Zündapp engine, located in the center of the vehicle.

The explosive charge was increased to 75 Kg. The improved range [12 Km] and cross-country ability made this version of the "Goliath" an effective weapon and a greater number were built and employed.

Leichter Ladungsträger "Goliath" [Sd.Kfz. 303a]

WEIGHT:	803 lb. [365 kg]
CREW:	— remote control
MOTOR:	1 X Zündapp 703 ccm, 2 cylinder 7.5 PS [HP] air-cooled
SPEED:	11 Km/h
RANGE:	Road — 12 Km, cross country — 7 Km.
CAPACITY:	6 liters [1.5 Imp. Gal.]
LENGTH:	162 cm.
WIDTH:	84 cm.
HEIGHT:	60 cm.
MANUFACTURER:	Zündapp Werke, Nürnberg, Zachertz, Freystadt.

The Sprengpanzer "Goliath" was designed to disable pillboxes, fortifications and heavy tanks as encountered by the Germans in France in 1940. The 270 Kg (appr. 810 lb.) weapon was powered by 2 X 2.5 Kw electro engines, giving the vehicle a range of less than one mile at top speed of 10 Km/h.

Built by Borgward and Zündapp during 1942-43, the remote controlled vehicle carried a load of 60 Kg of explosives which would be remotely activated after reaching the target.

The limited range, poor cross-country ability and high vulnerability made this weapon very ineffective. Production was halted in late 1943.

The "Goliath" was operated by a two men crew and wheeled to the target area on a special trailer.

"Goliath" Sprengpanzer are readied for action during the Warsaw uprising in August 1944. The soldier in the foreground is armed with a Gewehr 41[M] rifle, a semi-automatic gas-operated weapon.

Leichter Ladungsträger [Sd. Kfz. 303a] rear view

Schwerer Ladungsträger [Sd.Kfz. 301] Ausf. BIVa

Production of the schw. Ladungsträger [heavy demolition carrier] began in 1940 by Borgward, Bremen. The Ausf. BIVa was actually a small one-man tank powered by a six cylinder, 49 PS Borgward truck engine. The unit had a range of over 200 Km [appr. 110 miles] and would carry a 500 Kg [1,100 lb.] demolition charge in a special armored [10 mm] container located in the front of the vehicle.

The engine was housed in the rear of the vehicle. A 123 litre [appr. 30 Imp. gal.] fuel tank gave the 3.5 ton vehicle a range of 212 Km road and 125 Km cross-country range.

The driver who was only lightly protected by 10 mm armor plates, would drive the Sd.Kfz. 301 to the target area and after switching the instruments to radio control, abandoned the vehicle which continued, guided by remote control [EP3 with UKE6] to the target. Upon arrival, the demolition charge would be released and the vehicle was retrieved. The 500 Kg charge would then explode and it was attempted to reuse the very expensive carrier which cost RM 28,000., the same price as a four wheeled armored scout car [1942 prices].

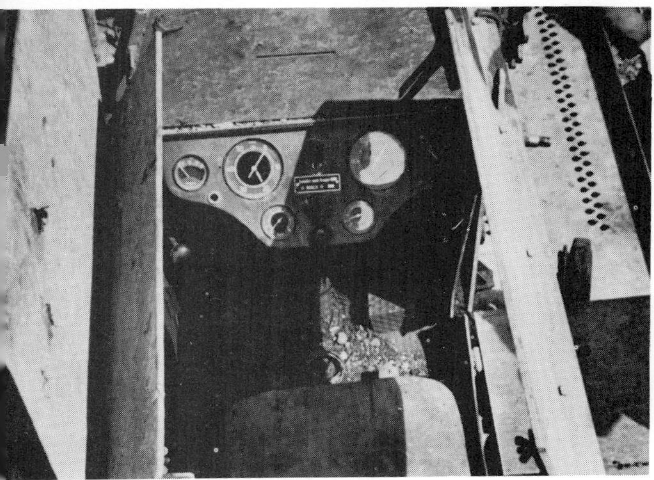

The driver's position showing the spartan instrument panel. Production of the schw. Ladungsträger stopped in 1943, after troop trials proved the application of the vehicle under existing conditions unrealistic, in addition to the numerous mechanical problems encountered in the field.

It was attempted to convert remaining chassis to other use, such as this version by mounting a 5 cm Pak 39 on a shortened chassis of the schw. Ladungsträger BIV.

To protect the gun and crew from bad weather, a standard truck canvas frame was available. Only a small number were converted and issued to Panzerjäger units.

An interesting conversion of the robust chassis of the Raupenschlepper Ost (RSO) primemover. This vehicle mounted a 7.5 cm Pak 40 anti-tank gun. The driver and gun crew were only protected against light infantry fire from the front and the gun had to be used from a hull down position.

10.5 cm K 18 auf Panzer Sfl.IVa

The Panzerjäger-Selbstfahrlafette [self-propelled tank hunter] was originally conceived as a weapon capable of defeating the heavy British and U.S. tanks which were to appear by 1942-43.

The vehicle was heavily armored in front, but only lightly armored on the sides and the rear. The fighting compartment was open on top. It is suggested that this weapon was able to defeat any Russian tank of its time at up to 2,000 meters. It is also reported that a number of T-34 fell victim to the accurate high velocity Krupp cannon before the unit was returned to Germany. The second Pz.Sfl.IV was destroyed by accidental ignition of the ammunition in the vehicle.

Armed with the Krupp designed 10.5 cm K 18 cannon, weighing over 4,000 lb. only two prototypes were built and sent to the Eastern Front for troop trials with the 3. Panzerdivision.

Below Rear view of the Panzerjäger with escape hatches on either side of the engine which doubled as loading hatches for the heavy 10.5 cm ammunition. A canvas was stored in the container on top of the rear compartment and could be rolled over the structure.

Schützenpanzer MARDER

The Schützenpanzer MARDER [MICV Marder] was developed and built by Thyssen-Henschel [formerly Rheinstahl] of Kassel, Germany. The mechanized infantry combat vehicle [MICV] is designed as a combination transport-combat vehicle for a group of seven Panzer Grenadiers, plus commander, driver and gunner, to operate in conjunction with the German Leopard main battle tank.

Schützenpanzer Neu MARDER [MICV MARDER]

WEIGHT:	28.2 t [28,200 Kg]
CREW:	10 [commander, gunner, driver and seven Pz. Grenadiers]
MOTOR:	MTU MB 833 Ea -500, 6 cylinder diesel engine 600 h.p. at 2,200 rpm.
SPEED:	75 Km [road] Range: 520 Km [road]
FUEL CAPACITY:	652 liters
LENGTH:	6.79 m
WIDTH:	3.24 m
HEIGHT:	2.86 m
V/OBSTACLES:	1.00 m
TRENCH:	2.50 m
GRADIENT:	60%
ARMAMENT:	1 X 20 mm Rheinstahl 202 automatic cannon elevation +65°, depr. -17°, 1 X 7.62 mm MG 3 co-axial, 1 X 7.62 mm MG 3 pintle mounted machine gun, 6 smoke discharger
DEVELOPED:	Thyssen-Henschel [1926 vehicles], Atlas MaK Kiel [875 vehicles].

Main Battle Tank Kampfpanzer Leopard 1

The 14.6 ton Schützenpanzer HS-30 was powered by a British 8 cylinder 236 HP Rolls-Royce B 81 MK 80 F engine. Armed with a 1 X 20 mm HS 820 gun in a revolving one man turret, the vehicle carried a crew of two (commander-gunner and driver) plus six Pz. Grenadiers. The top speed of 58 Km (road) compared with the Spz. Marder's 75 Km (32 m.p.h.) and max. range of 270 Km, Marder 520 Km (290 miles). Only the German Army operated the HS-30 and in 1971, over 1,800 were still in service.

The main battle tank of the new German Army (Bundeswehr) was the U.S. built M 47, and the role of the Schutzenpanzer was satisfactorily filled by the Spz.HS-30, developed by Hispano-Suiza of Switzerland. The vehicle was in production from 1958 until 1962 and performed well in speed, cross-country ability and range with the M 47. However, with the development of the Kpz. Leopard, it became necessary to provide a Mechanized Infantry Combat Vehicle able to match the performance of this new Kampfpanzer.

The Schützenpanzer MARDER is considered to be the most advanced MICV in the West. Speed, armor, cross-country ability and fire power make the MARDER a perfect match with the Kpz. LEOPARD.

The driver is sitting in a raised position and has a clear field of vision. A seat and hatch are located behind the driver serving several purposes. The position could be used by the commander or platoon leader, the gunner to aid the driver during night driving under non-combat conditions, accommodations for an observer for the artillery, a crew member of a disabled vehicle or for the driving instructor during training. A camouflage net is carried on the vehicle as standard equipment.

The first contracts for the construction of prototypes were awarded to Rheinstahl, Henschel and Mowag of Switzerland. The contract for the construction of 1926 Spz. MARDER was awarded to Rheinstahl [now Thyssen-Henschel] in 1969 and the first vehicle was handed over to the Bundeswehr on May 7, 1971.

The large rear ramp gives the crew an alternative to the roof hatches for embarkment under fire.

The MARDER has a sloped front hull which houses the Renk 194 automatic transmission. The "MLC-Schild" is a yellow round shield indicating the combat weight of the vehicle [29 ton for the MARDER]. This is important to know when negotiating bridges of various capacity in Europe during maneauvers and training.

Rear view of the Spz. MARDER. The exhaust grill is located to the extreme top right of the fighting compartment, while the louvers on both sides of the ramp are for air which is sucked in by two fans. Every Bundeswehr vehicle carries a license plate on the front and rear, plus a tactical symbol identifying the unit and vehicle number. This MICA belongs to the 2.Pz. Gren Btl. 53.

The Kettenschürzen [armor skirts] of the MARDER are the same as used for the LEOPARD and are made of steel braced rubber giving the tracks and road wheels additional protection, especially from hollow charges. The tracks are like the LEOPARD's of U.S. pattern with rubber pads.

The 20 mm Rh 202 cannon and 7.62 mm MG 3 machine gun are ideally suited as anti-aircraft weapons. The electro-hydraulic aiming device in the two-men turret allows the full cooperation between gunner and commander during action against low flying aircraft.

The MARDER crew during debriefing in the protection of their 29 ton vehicle. The armor skirts could easily be removed during training to avoid loss or damage of the equipment. The box on the left side of the turret houses a IR searchlight and the white light searchlight, with a range of 1,000 meters and 800 meters respectively.

The ten men crew of a Spz. MARDER is lined up in front of their vehicle. From front row left to right are: The commander, three Pz. Grenadiers and the driver, second row, the platoon-leader and four Pz. Grenadiers, each holding the very effective 9 mm Uzi sub machine gun.

The suspension of the MARDER consists of transversely mounted torsion bars with six road wheels on either side, with the first two and last two incorporating hydraulic shock absorbers. The drive sprocket is in front of the running gear, pulling the double pin type tracks over three top rollers.

A well camouflaged MARDER of the 3.Pz. Gren. L.Btl. 92 in the Lüneburger Heide. The IR searchlight box is removed from the turret.

An early version prototype of the MARDER during firing trials in Munster Lager, Northern Germany. The 20 mm Rh 202 cannon was chosen over the MK 20 mm Mauser Model 72 as the main weapon. 1,250 rounds of 20 mm ammunition plus 5,000 rds. 7.62 mm MG ammunition is carried in the vehicle.

The 29 ton MARDER could move with a top speed of 75 Km and the 652 ltr. fuel supply will allow a duration of 15 to 24 hours cross-country cruising time.

Panzer Grenadiers are firing their G 3 rifles from the open hatches. The Uzi sub-machine gun can be fired from within through the Kugelblenden of which two are installed on either side of the fighting compartment.

The early version of the Spz. MARDER was appr. 2 tons lighter than the newer type and had a simpler gun turret with no provision for the IR unit. The smoke candles are located on the lower turret and no armor skirts were fitted.

The powerful 600 H.P. engine is pushing the 27 ton MARDER through a river.

The driver and some of the crew of an early version of the MARDER.

The driver's position of the MARDER with the three periscopes giving the driver a 120° arc of vision.

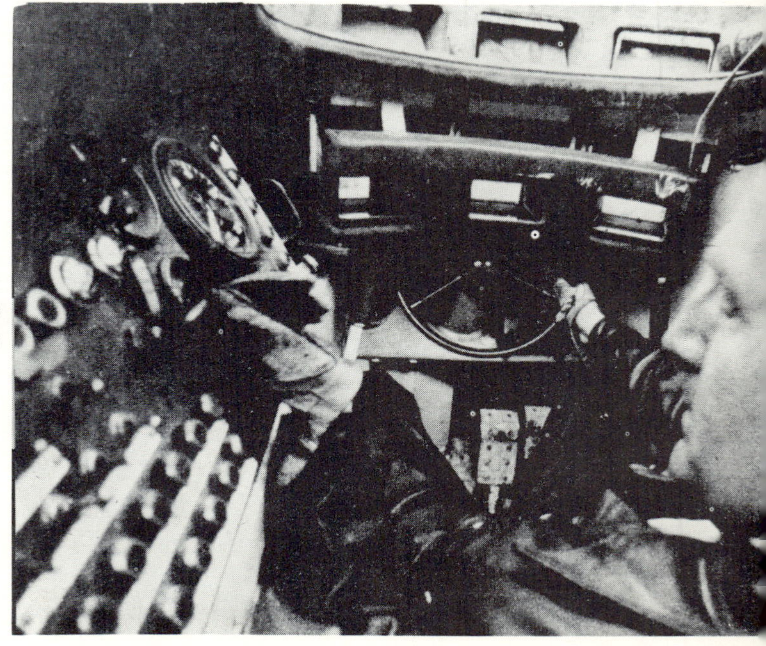

The Uzi sub-machine gun in firing position in the Kugelblende [ball-type mounting].

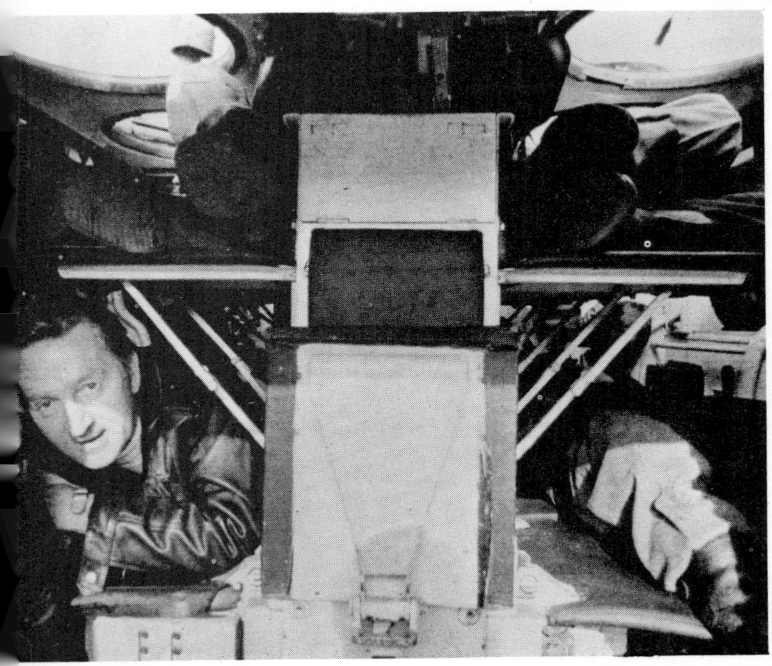

Four bunks could easily be converted from the seats by simply lowering the back rests. View is from the rear of the vehicle, with **the** four roof hatches open.

The powerplant layout and engine hatch of the ROLAND differ slightly from the original MARDER chassis. **A** ring type radiator is located in the rear and a supplementary generator for the electric power supply is installed in the engine compartment in front of the driver's position. The missile storage bins are open, able to hold five projectiles each.

The MARDER power-pack is lifted out by a Bergepanzer STANDARD. The speedy removal and installation of the massive MTU MB 833 Ea 500 engine with the transmission is emphasized by the use of multiple connector plugs for the electrical equipment. Quick release couplings for fuel and hydraulic pipes and all mechanical linkages have quick release pins and clips.

The chassis for the anti-aircraft system ROLAND prior to the installation of the missile unit.

Prototype of the ROLAND with the missile system in firing position. The weapon carrier stocks ten supersonic missiles. Targets flying as far as 6 Km from the ROLAND can be hit, detected by a sophisticated search radar system with a maximum range of 17 Km.

The ROLAND has a crew of three men and weighs 32.5 tons and is based on the modified chassis of the MICV MARDER.

Close-up of the suspension of the MARDER.

Rear view of the final version of the ROLAND.

The MARDER chassis is used as the basis as a carrier for a 120 mm or 160 mm mortar. Three hatches cover the fighting compartment housing the mortar bolted to the floor of the AFV. The five men crew includes the driver and gunner. The vehicle has a combat weight of 27.5 tons and a maximum range of 570 Km.

BERGEPANZER "PANTHER" [Sd.Kfz. 179]

The Bergepanzer "PANTHER" or Bergepanther filled the need for a fully tracked recovery vehicle able to retrieve the German heavy tanks, the Pz. V "Panther" and TIGER I. For air defence, either a 2 cm Flak or a machine gun could be installed in mountings attached at top center of the glacis plate next to the driver.

The winch controls were located in the radio-operator's position to the right of the driver.

The Bergepanther was based on the older Ausf. D. The turret was removed and a limited superstructure from wood and steel framing placed above the opening, housing tools, snatch-blocks and parts boxes. The winch filled the former fighting compartment to the level of the hull.

The earth spade in the lowered position. 500 ft. of 1 5/16″ cable was held by the winch, driven by a bevel-geared box powered from the main engine via the drive shaft.

A number of Bergepanther recovery vehicles had a huge spade attached to the rear of the superstructure. The spade could be lowered by the cable attached to a winch, giving the Bergepanther a solid base to exert its pulling capacity of 40 tons.

A 1½ ton jib was provided for lifting down equipment, spare engines, snatchblocks, etc. A chain of a 1½ ton capacity was attached with a built-in hook to a shackle on the boom.

Designed and converted by Demag, Weller/Ruhr, a total of 297 vehicles left the factory between 1944-1945.

The jib could be erected on either side of the rear superstructure.

View into the superstructure of a Bergepanther, holding two 23″ capstan drums, spring loaded roller and the main winch drum. A canvas cover protected the driver and crew from the elements.

The cable leading to the blade was guided through the vertical and horizontal roller pak, keeping the cable dead center to the winch and preventing entanglement during operation.

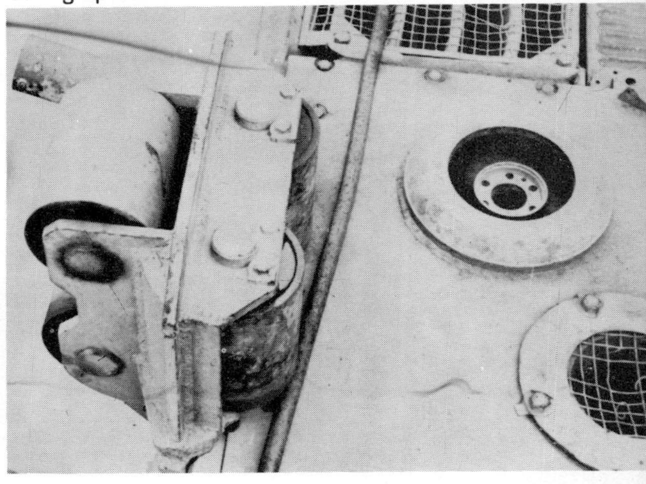

Rear view of a recovery tank with the earth blades removed.

Some Bergepanthers had a pedestal bolted on top of the glacis plate, a mounting for a 2 cm Flak. A Fliegergerät was sometimes welded on top of the ring to hold a MG 34 or 42 machine gun.

ARMOR SERIES

Each volume of this series consists of 52 pages, including four pages of full color drawings or action paintings of these vehicles. Descriptive text presents a view into the background and history of each vehicle and its development. Over 70 excellent photos, many never before published. Technical data rounds out the total coverage and provides a comprehensive look at that portion of the German armor story. 7½ x 11. Text and photographs by Walter J. Spielberger. Creative art by Uwe Feist.

Vol. 1

Vol. 2

THE GERMAN
PANZERS
from MarkI to MarkV "Panther"

Vol. 1 THE TIGER TANKS. Features all members of the Tiger Tank Family.
(p) $3.00 **ISBN 0-8168-2000-7**

Vol. 2 THE GERMAN PANZERS. From Mk. I to Mk. V "Panther." All other models of German tanks of the Second World War are covered in this volume. (p) $3.00 **ISBN 0-8168-2004-X**

Vol. 3 STURMARTILLERIE. From assault guns to Hunting Panther (Part I). Continues the development of the German armored force; featured are self-propelled guns and tank destroyers.
(p) $3.00 **ISBN 0-8168-2008-2**

Vol. 4 STURMARTILLERIE. Self-propelled guns and flak tanks (Part II). Concludes the chapter on those vehicles derived from basic tank chassis.
(p) $3.00 **ISBN 0-8168-2012-0**

Vol. 5 STRASSENPANZER. The full story of German four-, six- and eight-wheeled armored cars.
(p) $3.00 **ISBN 0-8168-2016-3**

Vol. 6 ARMOR ON THE EASTERN FRONT. An Armor Series Special shows German and Russian armored forces in action on the Russian front, 1941-1945. Side view drawings of the German Mark III F, Tiger (P) Elephant, Russian SU 85 and KW I(KVI). (p) $3.00 **ISBN 0-8168-2020-1**

Vol. 7 HALBKETTENFAHRZEUGE. Half-tracked vehicles. (p) $3.00 **ISBN 0-8168-2024-4**

Vol. 8 ARMOR IN THE WESTERN DESERT. Another pictorial special, featuring Allied and Axis armor in action in the North African campaign.
(p) $3.00 **ISBN 0-8168- 2028-7**

Vol. 9 SONDERPANZER. German special purpose and prototype vehicles. (p) $3.00 **ISBN 0-8168-2032-5**

Vol. 10 MILITARFAHRZEUGE. German softskinned vehicles. (p) $3.00 **ISBN 0-8168-2036-8**

Vol. 11 THE GERMAN 8 WHEEL SPAHPANZER "LUCHS" covers a wide range of interesting vehicles. The N.S.U. Kettenkrad, le.Zgkftwg. 1 t. Halftrack, 18 t "FAMO," Art. Beob. Pz. III, Stu. Pz. "Brummbar" VW Schwimmwagen, 4-wheel Pz.Sp.Wg., the new German Spah Panzer "Luchs" as the main feature.
(p) $3.95 **ISBN 0-8168-2040-6**

Vol. 3

Vol. 4

Vol. 5

Vol. 6

Vol. 7

Vol. 8

Vol. 9

Vol. 10

Vol. 11

DEUTSCHE PANZER 1917-1945. A collection of photos and data of the tanks, self-propelled guns and special purpose AFVs used by the German Army. 3½ pages of running gear close-ups of most every German AFV and a four page color section with 13 individual four color side-view, from the Panzer I-A through to the Tiger Elefant. 100 pages, size 8¼x11, 213 black and white photos depicting 97 different vehicles. Four pages with 13 color side-view, perfect binding. Price: $7.95.